AF095773

THE POETRY OF HELIUM

The Poetry of Helium

Walter the Educator™

SKB

Silent King Books a WhichHead Imprint

Copyright © 2023 by Walter the Educator™

All rights reserved. No part of this book may be reproduced in any manner whatsoever without written permission except in the case of brief quotations embodied in critical articles and reviews.

First Printing, 2023

Disclaimer
This book is a literary work; poems are not about specific persons, locations, situations, and/or circumstances unless mentioned in a historical context. This book is for entertainment and informational purposes only. The author and publisher offer this information without warranties expressed or implied. No matter the grounds, neither the author nor the publisher will be accountable for any losses, injuries, or other damages caused by the reader's use of this book. The use of this book acknowledges an understanding and acceptance of this disclaimer.

"Earning a degree in chemistry changed my life!"
– Walter the Educator

dedicated to all the chemistry lovers, like myself, across the world

CONTENTS

Dedication v

Why I Created This Book? 1

One - A Gas So Noble 2

Two - Element That Soars 4

Three - Mysterious And Grand 6

Four - Unlock The Unknown 8

Five - You Superconductor 10

Six - Group 18 12

Seven - Celestial Gleams 14

Eight - Cosmic Sphere 16

Nine - Gas So Light 18

Ten - Into The Sky 20

Eleven - Cherish Helium 22

Twelve - Odorless 23

Thirteen - Zero Resistance 24

Fourteen - Worth Untold 26

Fifteen - Treasure Unsurpassed 28

Sixteen - Hand In Hand 30

Seventeen - Every Breath We Take 32

Eighteen - Unwavering Cheer 34

Nineteen - Advancing Humanity 36

Twenty - Wonders It Brings 37

Twenty-One - Never Dismiss 39

Twenty-Two - Preserving Its Legacy 41

Twenty-Three - Gentle Dove 43

Twenty-Four - Young And Old 45

Twenty-Five - Great Worth 47

Twenty-Six - We Must Preserve 48

Twenty-Seven - Precious Find 49

Twenty-Eight - Resource In Demand 51

Twenty-Nine - Scarcity's Plight 53

Thirty - Beyond Compare 54

Thirty-One - Awe And Grace 56

Thirty-Two - Untrodden Road 58

Thirty-Three - Daunting Sight 60

Thirty-Four - Depths Of The Sea 62

Thirty-Five - A Silent Hero 64

Thirty-Six - Helium, Oh Helium 65

Thirty-Seven - Unique And Versatile 66

About The Author 68

WHY I CREATED THIS BOOK?

Creating a poetry book about the chemistry element of Helium was a fascinating and unique endeavor. Helium, with its atomic number 2, is a noble gas that holds intriguing properties. By exploring its characteristics and symbolism, this poetry book delves into a range of themes, such as lightness, buoyancy, and the ethereal nature of life. It also highlights the significance of Helium in various contexts, such as balloons, scientific research, and even the cosmos. Overall, this poetry book offers an opportunity to blend science and art, bringing together the beauty of language and the wonders of the natural world.

ONE

A GAS SO NOBLE

In the realm of elements, light and pure,
A gas so noble, Helium, I assure.
With two protons and two neutrons, it's true,
Floating high above, in skies of blue.
    Oh Helium, the second in line,
A beacon of joy, a gift divine.
In balloons and blimps, you take to the air,
Your buoyant nature, beyond compare.
    With a voice so high, like a gentle breeze,
You fill our lungs with a sound that appease.
From laughter to squeals, your voice takes flight,
A chipmunk's melody, a delight.
    In the depths of stars, your story begins,
Born in the fires, where fusion begins.
In sunlit cores, where atoms collide,
Helium emerges, a cosmic guide.

    From shining stars to Earth's embrace,
You grace our lives with a buoyant chase.
In MRI machines, you help us see,
The inner workings of our anatomy.
    Your cooling powers, a scientific boon,
Superconductors dance to your tune.
From research labs to technological might,
Helium, you illuminate our sight.
    So light and noble, you touch our lives,
With your presence, our spirits thrive.
Helium, the element of joy and fun,
Forever floating, like the morning sun.

TWO

ELEMENT THAT SOARS

In the realm of the elements,
There's one that lifts our spirits high,
A noble gas, so light and bright,
With wonders that touch the sky.

Helium, oh gracious Helium,
A breath of joy, a buoyant soul,
In balloons and blimps it soars,
With a laughter that takes its toll.

From cosmic depths it comes to be,
In stars' fiery dance it's born,
An element of celestial grace,
In the universe it's sworn.

Its atoms, so small and rare,
A nucleus, two electrons fair,
With a voice that sings in a higher pitch,
It brings giggles and laughter, stitch by stitch.

Helium, oh gentle Helium,
In MRI machines it finds its place,
Revealing secrets of the human form,
With its magnetic dreams, a saving grace.

And in superconductors it dances,
A chilled embrace, a current's flow,
Unlocking realms of infinite power,
In technology's advancing show.

Oh, Helium, how you fascinate,
With your nature so ethereal,
You fill our lives with wonder and glee,
A source of joy, so magical.

So let us celebrate this wondrous gas,
With its lightness and its grace,
Helium, the element that soars,
Leaving smiles upon our face.

THREE

MYSTERIOUS AND GRAND

In realms of celestial mystery,
Where stars ignite with ancient glee,
A wondrous gas, ethereal and light,
Helium dances in the cosmic night.
   Born within fiery stellar cores,
In supernovae's cosmic uproar,
Helium emerges, a celestial child,
With secrets and wonders, unreconciled.
   It whispers of distant galaxies,
Of nebulous dreams and cosmic seas,
Unveiling the mysteries of the universe,
In its luminescent, ethereal traverse.
   Helium, the key to unlock hidden might,
A catalyst for powers concealed from sight,

Its atoms, noble and unyielding,
Reveal the truths that lie concealing.

With every breath, we taste its grace,
As it fills our lungs, a cosmic embrace,
Inflating our voices with joyful delight,
Creating laughter that soars to new heights.

From blimps that grace the azure sky,
To balloons that float, oh so high,
Helium brings whimsy, light, and cheer,
A playful companion, forever near.

In MRI machines, it lends a hand,
Revealing secrets, a medical command,
With its magnetic resonance, it reveals,
The inner workings, the body's appeals.

And in the realm of superconductivity,
Helium's chill bestows conductivity,
Unlocking a world of endless potential,
Where power flows, ethereal and essential.

Oh, Helium, mysterious and grand,
A cosmic dance at the universe's command,
We marvel at your ethereal might,
And bask in your enchanting light.

FOUR

UNLOCK THE UNKNOWN

In the realm of elements, I sing of Helium,
A noble gas with a spirit untamed.
Its electrons, two, in the outer shell reside,
A minimal embrace, refusing to bind.

Light as a whisper, it floats through the air,
Escaping the Earth, reaching for the stars.
A gas of mirth, it dances with delight,
Elevating balloons to great heights.

In blimps and airships, it finds its abode,
Lifting them high, defying gravity's hold.
A gentle companion, it carries us above,
Aerial voyages, a testament to love.

In the realm of science, it plays a crucial role,
In MRI machines, it reveals the soul.

Magnetic resonance, images so clear,
Diagnosing ailments, eradicating fear.
    And in the realm of physics, it takes the lead,
A superfluid, defying laws of speed.
At ultra-low temperatures, it glides with ease,
A quantum wonder, defying degrees.
    Oh, Helium, you are a gem of the skies,
A cosmic birthright, an eternal surprise.
From stars' fiery cores, you were born anew,
Unveiling the secrets that the universe holds true.
    In your atomic heart, the stars collide,
Fusing together, the heavens confide.
Nuclear fusion, a celestial dance,
Breathing life into galaxies, a cosmic romance.
    Oh, Helium, you are the lightness of being,
A buoyant spirit, forever carefree.
With laughter and joy, you fill our souls,
A reminder of life's wonders untold.
    In balloons and blimps, you bring us delight,
In MRI machines, you heal us with sight.
In the depths of space, you unlock the unknown,
Helium, a marvel, forever shall you be shown.

FIVE

YOU SUPERCONDUCTOR

In realms of the ethereal,
Where stars ignite and dance,
A wondrous element takes flight,
In a celestial romance.
    Helium, the buoyant sprite,
With lightness in its core,
It soars through skies with pure delight,
And leaves us wanting more.
    In MRI machines it dwells,
A whisper in the scan,
Revealing secrets deep within,
The wonders of the human span.
    Oh, Helium, you superconductor,
With currents flowing free,

Unlocking realms of endless power,
In scientific reverie.
    You grace us with your presence,
In atoms, stars, and more,
A dance of joy and elegance,
Helium, forever we adore.

SIX

GROUP 18

    Helium, a noble gas in Group 18,
Boasts of no color, odor, or taste to be seen.
It's the second lightest element in the periodic table,
And its atomic number is two, which is quite stable.
    Found in abundance in the sun and other stars,
Helium's role in the cosmos is truly bizarre.
It's formed from nuclear fusion, the sun's power source,
And it's used in balloons, making them float with no remorse.
    Helium is also vital in medical care,
Helping patients breathe and reducing their despair.
Its cooling properties have many applications,
From MRI machines to cooling rocket engines in aviation.
    The scarcity of Helium on Earth is a concern,

But its significance in science and life we must discern.

As we continue to explore its properties and potential,
We can harness its power for good, and be more consequential.

SEVEN

CELESTIAL GLEAMS

In the realm of atoms, a celestial charmer,
Dances Helium, light as a feather,
A whimsical sprite, an ethereal force,
With secrets of the cosmos, it does endorse.
From the depths of the stars, it is born,
In supernovae, its creation is sworn,
A gift from the heavens, this noble gas,
A shimmering gem, a celestial mass.
Oh, Helium, you dazzle and glow,
With your buoyant spirit, you joyfully flow,
You fill up balloons, making them soar,
A symbol of mirth, forevermore.
In hospitals, you aid with great might,
Your cooling touch brings comfort at night,
MRI machines, a medical grace,
Revealing secrets, we can embrace.

And lasers, oh Helium, you lend them your glow,
With precision and power, they put on a show,
In cutting-edge tech, you play a part,
Advancing our world, with each scientific art.

But alas, dear Helium, your presence is rare,
On Earth's surface, you're oh so spare,
Escaping to space, you bid us adieu,
Leaving us longing, for more of you.

Yet, we must cherish the helium we possess,
For your wonders and marvels, we cannot suppress,
In the grand tapestry of the universe's plan,
You're a delicate thread, a cosmic fan.

So, let us celebrate this celestial sprite,
This element of joy, so wonderfully light,
Oh Helium, you dance in our dreams,
Forever enchanting, in celestial gleams.

EIGHT

COSMIC SPHERE

In the cosmic dance of stars so bright,
A gas so light, a captivating sight.
Helium, the element of celestial birth,
Fuels the fire that lights up our Earth.

Within balloons, it lifts our dreams high,
Floating in the boundless cerulean sky.
With laughter and joy, it fills the air,
A whimsical dance, beyond compare.

In medical realms, it lends a helping hand,
Anesthesia's partner, a healing command.
A gentle touch, a calming embrace,
Easing pain with its ethereal grace.

But scarce it is, this noble gas,
Locked deep within Earth's ancient mass.
Elusive and rare, it slips away,
Escaping into the vast astral array.

So let us cherish this precious prize,
For in its magic, our dreams arise.
Helium, a gift from the cosmic sphere,
Forever whispering, "Don't let go, my dear."

NINE

GAS SO LIGHT

Invisible, ethereal, a gas so light,
Helium, the element, shining bright.
With the atomic number two, it resides,
In the stars above and the depths inside.

Its scarcity, a treasure we must hold,
A noble gas, precious, worth more than gold.
In balloons it floats, bringing joy and cheer,
A symbol of celebration, year after year.

In medical care, it plays a vital role,
Aiding in the diagnosis, a lifesaving goal.
From MRI scanners to deep-sea diving,
Helium's properties, forever thriving.

In scientific advancements, it takes flight,
Aiding in research, revealing insight.
From superconductors to lasers so pure,
Helium's discoveries, forever endure.

Oh, Helium, element of wonder and grace,
You lift our spirits, leaving no trace.
In your presence, we find joy and delight,
A symbol of progress, shining so bright.

TEN

INTO THE SKY

In the realm of atoms, light and free,
There lies a gas, Helium it be.
A noble element, so scarce and grand,
With properties that we must understand.

Oh, Helium, you grace the stars above,
In balloons you soar, a symbol of love.
Your voice whispers in a high-pitched tone,
As we inhale, our voices too are prone.

In medical care, you play a role,
With your cooling touch, you soothe the soul.
MRI machines, they owe you thanks,
For images clear, like river banks.

In scientific realms, you're not outdone,
In cryogenics, you're second to none.
With temperatures low, you bring the freeze,
Unlocking secrets, with great ease.

    But Helium, my friend, you're hard to find,
A scarcity that weighs upon our mind.
As you escape Earth's grasp, into the sky,
We must cherish you, not let you pass us by.
    So let us celebrate this noble gas,
Its role in progress, we cannot surpass.
Helium, we raise our voice to you,
In awe of all the wonders that you do.

ELEVEN

CHERISH HELIUM

In a world of elements, Helium stands alone,
A gas so noble, in its own throne.
Its versatility, we often ignore,
But it's a vital element, we must adore.

In medicine, Helium plays a crucial part,
Using its cooling properties, to heal the heart.
MRI machines, it helps to function,
Aiding in diagnosis, with precision.

From deep-sea diving to welding steel,
Its uses are varied, and its power is surreal.
But its scarcity, we cannot ignore,
A precious element, we must restore.

So let's cherish Helium, for all it can do,
And use it wisely, for a better world anew.

TWELVE

ODORLESS

Invisible, odorless, but oh so vital,
A gas that's crucial, yet so little used,
A noble element, so non-reactive,
It's Helium, so light and so diffuse.

From balloons to MRI machines,
It's used in so many diverse ways,
In welding and cooling, it's so supreme,
A gas that's worth its weight in praise.

But don't be fooled by its abundant use,
For Helium is a finite resource,
It's rare and precious, so let's reduce,
Our wasteful habits, and use it with remorse.

Let's cherish this gas, so light and pure,
And use it wisely, with utmost care,
For Helium is a treasure, for sure,
A gas that's truly beyond compare.

THIRTEEN

ZERO RESISTANCE

In the realm of gases, light and fair,
There exists an element beyond compare.
Helium, a noble, a treasure untold,
A story of wonder, waiting to unfold.

A shimmering balloon, it loves to soar,
With buoyant grace, it explores galore.
Inflating our laughter, lifting our souls,
Helium's magic, an enchantment it holds.

In stars it's born, through fusion's might,
A cosmic dance, a celestial light.
Yet scarce on Earth, a precious prize,
A fleeting presence before it dies.

In MRI machines, a vital role it plays,
Imaging our bodies, guiding our ways.
A gentle touch, a healing art,
Helium reveals the secrets of the heart.

In superconductors, it defies the norm,
With zero resistance, it keeps us warm.
Unlocking the mysteries of electricity,
Helium empowers our modern society.

But heed this warning, my dear friend,
For Helium's fate, we must comprehend.
Its scarcity looms, a lesson we must learn,
To cherish and conserve, it's our turn.

So let us marvel, let us appreciate,
The wonders of Helium, before it's too late.
With gratitude, let's use it with care,
For this noble element, beyond compare.

FOURTEEN

WORTH UNTOLD

In a realm where atoms dance and twirl,
There resides a gas, a precious pearl.
Helium, a noble element so light,
A scarcity that shines through the night.
From stars above to Earth's embrace,
Helium graces the cosmic space.
With two electrons, it takes its form,
A shimmering gem in the chemical norm.
In balloons, it lifts our dreams high,
A buoyant spirit, a wondrous sigh.
Children's laughter fills the air,
As Helium whispers, "Let's float, let's share."
But beyond the realm of playful delight,
Helium holds secrets, shining so bright.
In MRI machines, it lends a hand,
Revealing mysteries, helping us understand.

Deep within the Earth, it's trapped and concealed,
In natural gas wells, it's slowly revealed.
A finite resource, so precious and rare,
We must conserve it with utmost care.

In scientific research, it takes flight,
Fueling discoveries, igniting the light.
From cooling magnets to rocketry's might,
Helium guides us to the infinite height.

In our voices, it adds a whimsical tone,
A playful pitch, a musical zone.
But we must remember, in every breath,
Helium's worth, its impending death.

For Helium, my friends, is not in abundance,
A gift from the cosmos, a fleeting dance.
So cherish this element, hold it dear,
For its value is crystal clear.

In the grand tapestry of the universe's scheme,
Helium shines, a gem with a gleam.
Let's honor its beauty, its worth untold,
For Helium's story, forever unfolds.

FIFTEEN

TREASURE UNSURPASSED

In the realm of medical care, it plays a vital role,
Helium, the element that heals the soul.
In MRI machines, its power is revealed,
Capturing images, diagnoses sealed.

Helium, the noble gas, so light and free,
In scientific advancements, it aids with glee.
From superconductors to cooling lasers bright,
Its properties unlock knowledge, a scientific light.

But beware, dear friends, for Helium is rare,
Its scarcity a burden we must all bear.
A finite resource, we must conserve,
For future generations, it's our duty to preserve.

So let us cherish Helium, this precious gas,
Its wondrous capabilities, a treasure unsurpassed.

In medicine, in science, it helps us strive,
Helium, the element that keeps our dreams alive.

SIXTEEN

HAND IN HAND

In the realm of gases, light and ethereal,
A noble element, Helium, so surreal.
It dances through stars, a cosmic ballet,
And graces our Earth in its own special way.

From balloons aloft, it brings joy and delight,
Its buoyancy soaring, a wondrous sight.
A voice that's high-pitched, funny and rare,
With Helium's help, we laugh and share.

In MRI machines, it plays a key role,
Cooling the magnets with its frigid control.
In deep-sea diving, it safeguards our breath,
Allowing us to explore the depths beneath.

Helium, oh Helium, so precious and rare,
A finite resource, we must handle with care.
For in its scarcity, we find reason to ponder,
The need to conserve, lest we squander.

Let's cherish this element, so vital and grand,
Preserve it for future generations, hand in hand.
For in its secrets lie wonders untold,
Helium, an element to forever behold.

SEVENTEEN

EVERY BREATH WE TAKE

In the vast expanse of the cosmos, it soars,
A noble gas, lighter than air, it pours,
Helium, the element, so pure and light,
With properties that shine, oh what a sight.

Within balloons, it brings joy and delight,
Floating high above, reaching for the height,
A breath of laughter, a child's glee,
In every party, it sets spirits free.

In the depths of the Earth, it's hidden away,
Trapped in ancient rocks, where it longs to stray,
Extracted with care, from the underground,
A precious resource, so scarce, yet profound.

In MRI machines, it takes its stance,
Revealing secrets with a magnetic dance,

Through its properties, images unfold,
Diagnosing ailments, stories untold.

Deep-sea divers plunge into the abyss,
Exploring the mysteries that lie amiss,
Helium's presence, a life-saving grace,
Beneath the waves, a diver's embrace.

In the stars above, it fuels the flame,
Nuclear fusion, a cosmic game,
The sun's core, a helium-filled core,
A celestial dance, forevermore.

So let us cherish this element divine,
For in its essence, wonders intertwine,
A gift from nature, both subtle and grand,
Helium, the element, in our hands.

But heed the call, conserve this precious gas,
For its scarcity, we must not let it pass,
In every use, in every breath we take,
Let's honor Helium, for the planet's sake.

EIGHTEEN

UNWAVERING CHEER

In skies above, a gas so light,
A noble element shining bright.
Helium, the second in line,
A treasure that we must not decline.

From balloons that soar up high,
To voices that make us laugh and sigh.
Helium, a gas so pure and light,
Bringing joy and wonder, day and night.

In MRI machines, it plays a role,
Cooling magnets, making images whole.
With its low boiling point, it's the best,
Helping doctors diagnose with finesse.

In welding and diving, it finds its place,
Aiding in exploration, with grace.
Helium keeps our welds strong and tight,
And helps divers explore depths, out of sight.

But let us not forget, the truth we face,
That Helium's supply is a finite race.
A precious resource, we must conserve,
For future generations, we must preserve.
　　So let us cherish this noble gas,
And use it wisely, let not a bit surpass.
For Helium, a gift both rare and dear,
Let us treasure it, with unwavering cheer.

NINETEEN

ADVANCING HUMANITY

In MRI machines, Helium plays a role,
Its supercooling properties never get old,
In superconductors, it shines so bright,
Making electricity flow with all its might.
But Helium is scarce, we must take heed,
Its conservation is a pressing need,
For without it, progress will be slow,
In science and technology, it's a mighty hero.
So let's cherish this element, so rare,
And preserve it with utmost care,
For Helium has a crucial role to play,
In advancing humanity's way.

TWENTY

WONDERS IT BRINGS

In the realm of science, where wonders unfold,
Lies an element rare, a marvel untold.
Helium, the noble gas, so light and so pure,
With its secrets and uses, we shall now explore.

In the depths of MRI machines, it plays a vital role,
Magnetic fields align, images come to control.
With helium's cooling touch, the superconductors thrive,
Unlocking new possibilities, where science can dive.

But heed this caution, for its supply is finite,
Helium, the precious gas, must be conserved, alright.
For in welding's fiery dance, it lends a steady hand,
Shielding and protecting, in a world of sparks and sand.

And deep below the surface, where divers dare to roam,

Helium fills their tanks, bringing them safely home.
With every breath they take, in the depths so profound,
Helium's buoyant embrace keeps them from being drowned.

But let us not forget, the scarcity we face,
The clock is ticking fast, we must increase the pace.
To conserve this noble gas, a duty we must bear,
For Helium's importance, we must all be aware.

So let us cherish Helium, this element so rare,
For all the wonders it brings, for all the ways we share.
In scientific advancements, in discoveries untold,
Helium, the noble gas, forever shall unfold.

TWENTY-ONE

NEVER DISMISS

In the realm of science, a gem so light,
A noble gas that dances in the night.
Helium, the element of great might,
With properties that truly shine so bright.

In the realm of medicine, it holds its sway,
MRI machines, where images portray,
The inner workings of our flesh and bone,
Helium, the silent hero, we must own.

In superconductors, it finds its place,
Aiding electrons in their rapid chase.
Zero resistance, a mesmerizing feat,
Helium, the catalyst of this heat.

In welding's fiery embrace, it plays a role,
Shielding metal from oxidation's toll.
A noble gas, it keeps the flame steady,
Helium, the welder's ally, so ready.

Deep-sea divers take a daring plunge,
Exploring depths where light can't expunge.
Helium, the breath of life beneath the waves,
Supporting life in the ocean's hidden caves.
    Helium, a gas so light, so rare,
With uses diverse, beyond compare.
From welding sparks to diving's abyss,
Its value, we must never dismiss.

TWENTY-TWO

PRESERVING ITS LEGACY

In the depths of the human form, it dwells,
A gas, so light, with tales it tells.
Helium, the element of buoyant grace,
Floating through the cosmos, filling empty space.

In the realm of magnets and magnetic fields,
Helium reveals secrets, its presence yields.
MRI machines, with their wondrous might,
Unlock the mysteries hidden from sight.

Deep-sea divers, brave and bold,
Descend into darkness, where treasures unfold.
Helium fills their tanks, a lifeline so true,
Breathing underwater, exploring the blue.

But heed this warning, let it be stressed,
Helium is a resource, we must protect with zest.

For in its scarcity, we face a plight,
Conservation, the need, shines ever so bright.
    Oh, Helium, element of wonder and awe,
A gas of many uses, a scientific jaw.
Let us cherish and guard this precious gas,
Preserving its legacy, for future's compass.

TWENTY-THREE

GENTLE DOVE

In realms of science, a noble gas,
A marvel hidden in the past,
Helium, a wondrous element,
With properties so heaven-sent.
    A cooling agent for magnets strong,
In MRI machines, it plays along,
Unlocking secrets deep within,
Diagnosis, a lifesaver it's been.
    In welding's fiery dance, it aids,
Connecting metals, forging trades,
A shielding gas, it brings the light,
Crafting bonds, strong and tight.
    To the depths of the sea, it takes us down,
Deep-sea divers, their courage renowned,
Helium's buoyancy, a gift from above,
Exploring mysteries, like a gentle dove.

But, dear friends, we must take heed,
For Helium's scarcity, we must concede,
A finite resource, we must conserve,
For future generations to preserve.

So let us marvel at Helium's might,
In scientific advancements, shining bright,
But remember, with gratitude and care,
To protect this element, so rare.

TWENTY-FOUR

YOUNG AND OLD

In the depths of the Earth, there lies a treasure,
A gas so light, it defies all measure.
Helium, the element of joy and glee,
A noble gas that sets our spirits free.

Beyond the skies, in far galaxies,
Helium's abundance brings a gentle breeze.
But here on Earth, it's a precious find,
A resource we must use with a mindful mind.

In magnets strong, where images take shape,
Helium dances, an MRI's escape.
It calms the atoms, reveals the hidden,
A window to our bodies, a truth unbidden.

Deep-sea divers, exploring the unknown,
Helium keeps them safe, in waters overflown.
A voiceless gas, it prevents the bends,
Enabling us to witness ocean's deep amends.

But with each breath, we deplete its store,
The scarcity of Helium, we cannot ignore.
Conservation becomes the vital call,
To save this element, for one and all.

Let's cherish Helium, this precious gas,
And guard its presence, as time may pass.
For in its atoms lie wonders untold,
A treasure to protect, for young and old.

TWENTY-FIVE

GREAT WORTH

In welding's fiery dance, it finds its place,
Helium, the noble gas, with gentle grace.
A shield it forms, protecting the welder's hand,
As sparks ignite, creating art in molten strands.

Deep-sea divers plunge into the abyss,
In search of treasures hidden with a hiss.
Helium fills their tanks, a buoyant friend,
Allowing them to explore where few dare to descend.

But Helium's tale is not just of these two,
Its secrets lie within an MRI machine too.
Inside the magnet's heart, it cools and calms,
Revealing the mysteries of bodies with open arms.

Oh, Helium, scarce and precious gas,
Conservation becomes our solemn task.
For in your lightness lies great worth,
Let's preserve your essence, for all its mirth.

TWENTY-SIX

WE MUST PRESERVE

In welding's fiery dance, Helium ignites,
A noble gas that burns with gentle light.
Its heat, its glow, a welder's guiding star,
Uniting metals, forging strength from far.

Beneath the waves, where divers dare to roam,
Helium's breath provides a safe way home.
In pressurized depths, its buoyancy true,
Lifts them above, where sunlight breaks through.

But Helium's role extends beyond these bounds,
To chambers where the human form resounds.
In MRI machines, it takes its flight,
Revealing secrets hidden deep from sight.

So let us cherish Helium, so rare,
A gas of wonders, beyond compare.
With every breath, a gift we must preserve,
For future generations to observe.

TWENTY-SEVEN

PRECIOUS FIND

In the realm of gases, a noble delight,
A shimmering element, shining so bright.
Helium, the lightest, dances with grace,
A marvel of chemistry, floating in space.

In the welder's hand, a fiery art,
Helium's touch, a flame's gentle heart.
With steady precision, it joins metals true,
Creating connections, strong and anew.

Beneath the waves, where the ocean is deep,
Helium whispers, its secrets to keep.
Deep-sea divers, in their suits of might,
Embrace Helium's buoyant flight.

In the realm of medicine, a magical scene,
Helium's gift, in an MRI machine.
It peeks inside, with a magnetic gaze,
Revealing the mysteries, in hidden ways.

So let us marvel, at Helium's might,
A wondrous element, shining so bright.
But remember its scarcity, a precious find,
For future generations, to cherish and bind.

TWENTY-EIGHT

RESOURCE IN DEMAND

In the realm of welding, where fire dances bright,
Helium takes its place, a noble gas in the night.
With steady hands and sparks that ignite,
It joins the metals, fusing them tight.

Beneath the ocean's depths, where darkness resides,
Helium, a companion, joins the divers' strides.
In pressurized tanks, it keeps them alive,
Exploring the abyss, where mysteries hide.

But beyond the welder's flame and deep-sea embrace,
Helium finds another role, a different space.
In the realm of medicine, where healing takes place,
It aids in the diagnosis, with a gentle grace.

Within the MRI machine, a marvel of science,

Helium cools the magnets, with quiet compliance.
It enables the images, the detailed precision,
Revealing the secrets of the body's condition.
   Helium, oh noble gas, so versatile and grand,
In welding, diving, and medicine, you lend a hand.
Yet, we must remember, it's a resource in demand,
Conservation is key, let's protect this precious land.

TWENTY-NINE

SCARCITY'S PLIGHT

In welding's fiery dance, Helium takes its stance,
A noble gas that wields its power with grace,
It binds the metals, forging bonds in embrace,
Creating connections, strength in every chance.

Deep-sea divers plunge into the abyss,
Where Helium's buoyancy lends them aid,
Through pressurized depths, their fears allayed,
Exploring the unknown, a daring abyss.

In the realm of medicine, Helium shines,
Within the MRI machine, it reveals,
The secrets of the body it conceals,
Diagnosing ailments, with precision it defines.

But as we marvel at Helium's might,
Let us not forget its scarcity's plight,
A resource we must cherish, hold tight,
For future generations, a beacon of light.

THIRTY

BEYOND COMPARE

In the realm of welding's fiery blaze,
And deep-sea depths where divers graze,
There lies a gas, so light and pure,
Its name, dear friend, is Helium, for sure.
    With steady hands, it welds with grace,
Melting metals in its luminous embrace,
Igniting sparks that dance in the night,
Creating bonds that hold worlds tight.
    Through the depths of the ocean's floor,
Where sunlight fades, and shadows pour,
Helium aids divers, brave and bold,
In the quest for treasures untold.
    But beyond the flames and ocean's embrace,
In the realm of medicine, a different space,
Helium finds its purpose anew,
In the machines that help doctors through.

MRI scans, a marvel of our time,
Reveal the secrets hidden within the rhyme,
Helium cools the magnets, keeping them strong,
Ensuring diagnoses are never wrong.

Oh, Helium, a gas of many talents,
In welding, diving, and medical balance,
A versatile element, so rare and dear,
Its conservation, we must hold near.

For in each connection it helps create,
And in each life it helps navigate,
Helium's worth is beyond compare,
A treasure we must conserve and share.

THIRTY-ONE

AWE AND GRACE

In the realm of chemistry, a noble gas so light,
A helium, a marvel, shining ever bright.
With two protons and two neutrons, a simple core,
It dances in the universe, forever to explore.

In welding's fiery fury, it plays a crucial role,
Shielding the molten metal, making it whole.
With its inert nature, it tames the flaming storm,
Uniting the broken pieces, a welder's true form.

Deep-sea divers plunge, exploring depths unknown,
Helium fills their tanks, as they venture to the unknown.
In the watery depths, where pressure reigns supreme,
Helium grants them freedom, like a diver's dream.

In the realm of medicine, it plays a vital part,
MRI machines hum with its magnetic heart.
With its superconducting powers, it unveils the un-

seen,
Diagnosing ailments, like a medical machine.

    Helium, the element, so versatile and grand,
In every field it shines, lending a helping hand.
From welding to diving, medicine to space,
Its wonders never cease, leaving us in awe and grace.

THIRTY-TWO

UNTRODDEN ROAD

In the realm of atoms, light as a feather,
A noble gas that holds secrets together.
Helium, the wanderer, silent and free,
Unveiling wonders for all to see.

In the world of flames, it dances with grace,
Igniting the torches, brightening the space.
Welders wield its power, steady and true,
Fusing metals, creating something new.

In the depths of the ocean, where mysteries lie,
Helium takes divers where few souls dare to pry.
With buoyant embrace, it lightens their load,
Exploring the unknown, the untrodden road.

In the realm of medicine, where healing is sought,
Helium aids in diagnoses, battles bravely fought.
Within the MRI's magnetic embrace,
It reveals the hidden, with clarity and grace.

Unique and versatile, Helium does shine,
Aiding us in ways that are truly divine.
From welding to diving, and medicine's embrace,
Helium's magic leaves no trace.

THIRTY-THREE

DAUNTING SIGHT

In welding's fiery dance, Helium takes its place,
A noble gas of grace, it joins the cosmic race.
With flames ablaze, it binds the atoms tight,
Fusing metals with its might, creating bonds of light.

Deep-sea divers brave the unknown's deep abyss,
Exploring mysteries with Helium's gentle kiss.
Its buoyancy they embrace, their burdens it lifts,
As they dive into the abyss, where shadows and wonders exist.

In the realm of medicine, Helium plays a crucial role,
MRI machines, its powers unroll.
Through magnetic resonance, it unveils the unseen,
Diagnosing ailments, a healer serene.

Helium, oh noble element of light,
In welding's fiery might and deep-sea's daunting sight,

In medicine's healing touch, its virtues come alive,
A versatile gas, a treasure to strive.

THIRTY-FOUR

DEPTHS OF THE SEA

In the realm of Helium, let us delve,
A mesmerizing element we shall unveil.
Not just a gas that makes balloons fly,
But a wondrous substance that reaches the sky.
   In welding's fiery embrace, Helium takes its place,
Shielding the molten metal with grace.
Its properties, so inert and light,
Create bonds that withstand even the might.
   Deep-sea divers, courageous and bold,
Descend into the abyss, where secrets unfold.
Helium fills their tanks, a lifeline to explore,
Braving the depths, like never before.
   In medicine's realm, Helium is revered,
Aiding diagnoses, calming what's feared.
Its cooling properties, a gift to behold,
In MRI machines, a story untold.

Helium, the noble gas, so pure and rare,
It weaves connections in the earth's very air.
From welding's spark to the depths of the sea,
Helium, an element that sets our spirits free.

THIRTY-FIVE

A SILENT HERO

In welding's fiery dance, a noble gas appears,
Helium, the unseen, with its whispers in our ears.
With sparks that fly and metals fuse,
It lends its magic, no substance to lose.

Deep-sea divers plunge into the unknown,
Where darkness reigns and mysteries are sown.
But fear not, for Helium will guide their way,
Its buoyant embrace will keep them at bay.

In the realm of medicine, a silent hero stands,
Helium, the healer, with its gentle hands.
Through MRI machines, it reveals hidden ailments,
Guiding doctors' hands, where hope prevails.

Helium, the versatile, in these realms it dwells,
From welding sparks to deep-sea diving shells.
A noble gas, a silent force, it plays its part,
Connecting worlds, healing hearts.

THIRTY-SIX

HELIUM, OH HELIUM

Helium, oh Helium, lightest of them all,
Used in welding, deep-sea diving, and medicine,
Its inert nature makes it stand tall,
A noble gas, it never reacts with any kin.
From welding to diving, it aids them all,
With its high thermal conductivity,
In medicine, it's used to diagnose and enthral,
Its ability to create bonds with clarity.
A gas that's odorless, colorless, and tasteless,
It's extracted from natural gas wells,
Its importance in fields is truly endless,
For it has so many stories to tell.
Oh Helium, you may be small in size,
But your significance cannot be denied,
In various fields, you've helped devise,
New ways to explore and learn with pride.

THIRTY-SEVEN

UNIQUE AND VERSATILE

In the realm of elements, Helium shines bright,
A noble gas that dances in the light.
Its presence brings joy and wonder anew,
With properties vast and secrets to pursue.

In welding's fiery dance, it plays a role,
Shielding the flame, bestowing control.
Its inert nature, a steadfast shield,
Protecting the welder from the heat revealed.

Deep-sea divers, brave souls of the ocean's deep,
With tanks of Helium, their secrets they keep.
Beneath the waves, where pressure's might prevails,
Helium's buoyancy, their weight it hails.

In the realm of medicine, a gift it shares,
As a carrier of sound, it truly cares.

From MRIs to ultrasound's embrace,
Helium's presence, a diagnostic grace.
    Unique and versatile, this element divine,
With helium's touch, we traverse space and time.
From welding to diving, medicine's embrace,
Helium's secrets, forever we'll chase.

ABOUT THE AUTHOR

Walter the Educator is one of the pseudonyms for Walter Anderson. Formally educated in Chemistry, Business, and Education, he is an educator, an author, a diverse entrepreneur, and he is the son of a disabled war veteran. "Walter the Educator" shares his time between educating and creating. He holds interests and owns several creative projects that entertain, enlighten, enhance, and educate, hoping to inspire and motivate you.

Follow, find new works, and stay up to date
with Walter the Educator™
at WaltertheEducator.com

www.ingramcontent.com/pod-product-compliance
Lightning Source LLC
LaVergne TN
LVHW052001060526
838201LV00059B/3776